THE BOY, THE BIRD, AND THE TURBINE

By Seton LaSalle Faculty and Students

Emily Rosati Anthony DeCaria

Diego Flores/Cruz   Tyler Hill   Sethan-Jai Doan   Caroline Marston

DaVinci (Joshua Mellor)   Joseph Rouse   Taylor Weyrich

Illustrated by Sloane McCensky

Dedicated to the researchers of the American Wind Wildlife Institute

Text by Emily Rosati, Anthony DeCaria, Diego Flores/Cruz, Tyler Hill, Sethan-Jai Doan, Caroline Marston, DaVinci (Joshua Mellor), Taylor Weyrich, and Joseph Rouse, with Christine Real de Azua and Dr. Ellen Cavanaugh

Images by Sloane McCensky

Special thanks to Saint Francis University Institute for Energy for their support and wisdom.

ISBN 978-1-387-70240-4

Grow a Generation
Sewickley, PA 15143
www.growageneration.com

Any and all profits from the sale of this book benefit American Wind Wildlife Institute.

Once upon a time there was young boy named John who loved the woods and the Appalachian mountains of Western Pennsylvania.

John's mom worked for a local wind farm. She was a brave woman, unafraid to climb 150 meters with just a harness to do her tests and repairs on the wind turbine. She also liked the elevator ride up the 300 meter tall turbines.

Every day, John explored with his
binoculars and loved the wildlife,
particularly the wildlife that could fly.
One day while he was out exploring with
his mom, he met a red tailed hawk named
Beamer. Beamer told John all about his
life in Pennsylvania.

Beamer flew through the breezy mountains and rolling hills of Pennsylvania with all of his friends and family. Together, they flew with light breezes and strong gusts of wind. The sky was a wild and fascinating place for Beamer and his family.

One day, Beamer realized that something unfamiliar arrived at his home. Its appearance began to affect Beamer's way of life. He noticed his friends the bats would sometimes fly too close to the white tower's spinning blades. Some would fall.

Beamer became very sad. John could not bear to hear anymore of this story. He said, "We need to do something!"

Beamer and John approached the tall
white structure while it was not spinning
and shouted, "Why don't you pick on
someone your own size?!"

The tall white structure groaned and said, "I'm sorry. My name is Tad the Wind Turbine. I don't know how to stop hurting your friends while doing my job."

Beamer asked, "Well, what is your job?"

Tad shouted with authority, "I create sustainable energy! I battle climate change! I conserve water! I fight air pollution! I prevent acid rain!"

Tad explained, "Sustainable energy is a way to create clean electricity for years and years to come without hurting our environment. Sustainable energy, such as wind energy, is a type of energy that is able to last as long as the wind blows."

"Currently, fossil fuels need to be burned to make electricity. Burning causes air pollution. Air pollution causes acid rain which harms you and your friends, far more than running into turbines."

Beamer looked into the sky nervously.

"What is acid rain?" he asked.

Tad explained, "Acid rain is formed when compound chemicals like sulfur dioxide and nitrogen oxides are added to the air by burning fossil fuels. The chemicals combine with water in the clouds. When acid rain falls from the sky, it destroys plants and soil, makes water undrinkable, and even erodes buildings."

John added, "I learned about this in school. Acid rain causes aluminum to be released by the soil and kills the crayfish and fish that you like to eat from the stream."

Tad said, "I also help conserve water for wildlife and people. I keep water from being wasted when it is used for steam and for cooling in conventional fossil and nuclear power plants."

Despite his bragging, Tad seemed very upset that the important job, fighting climate change, was also harming the winged creatures. His blades tilted toward John and Beamer and he said gently, "What can I do?"

Wind energy is expanding quickly around the world. Wind power in the United States is a little more than 6% of all generated electrical energy power in the United States.

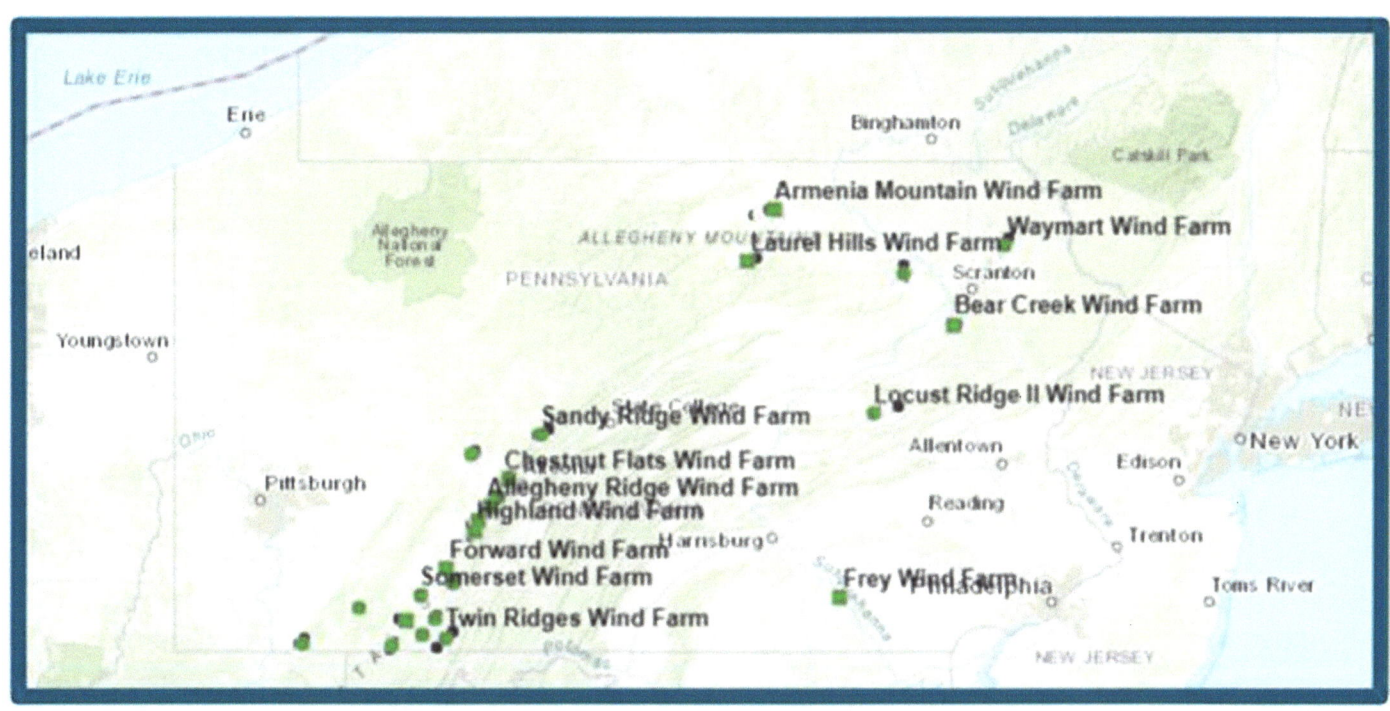

Look how much it has grown in Pennsylvania so far! This 2017 map represents 726 wind turbines on 25 wind farms. It was created by St. Francis University's Institute for Energy.

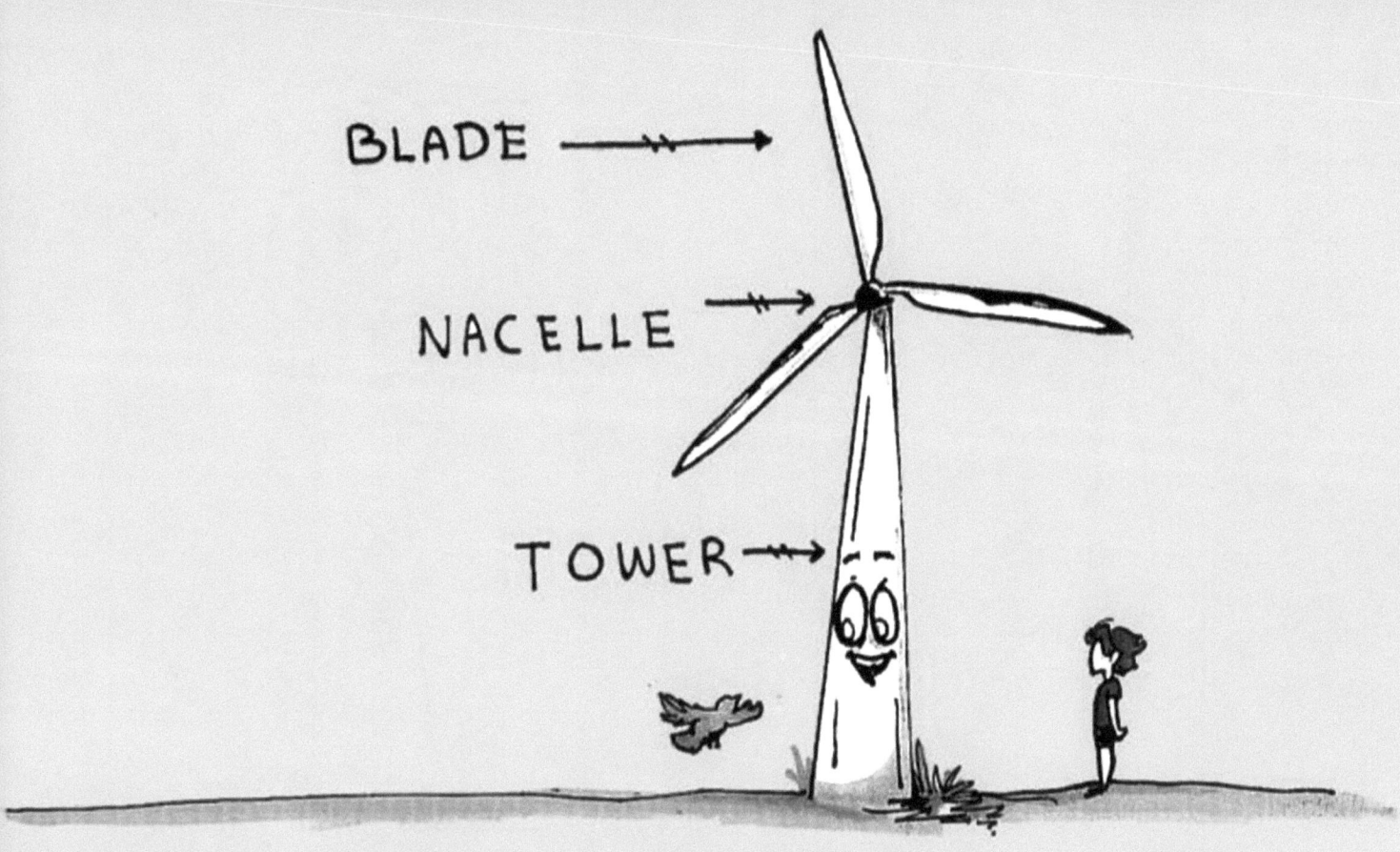

Tad was happy to explain how a turbine works. He started at the top. "Come on inside!"

Wind turns the blades which turn the generator inside the nacelle (the housing for the rotors, gear box, and generator).

The blades on a windmill are like the blades on a airplane. When air flows past the blade, a difference in pressure is created between the upper and lower blade surfaces. When the wind moves faster the blade goes faster.

My blades are connected to a gear box. Just like the gears on your bicycle, if you were riding up the hill (or the wind is not blowing fast), you would want the lowest gear so you don't put strain on your leg muscles or run out of breath.

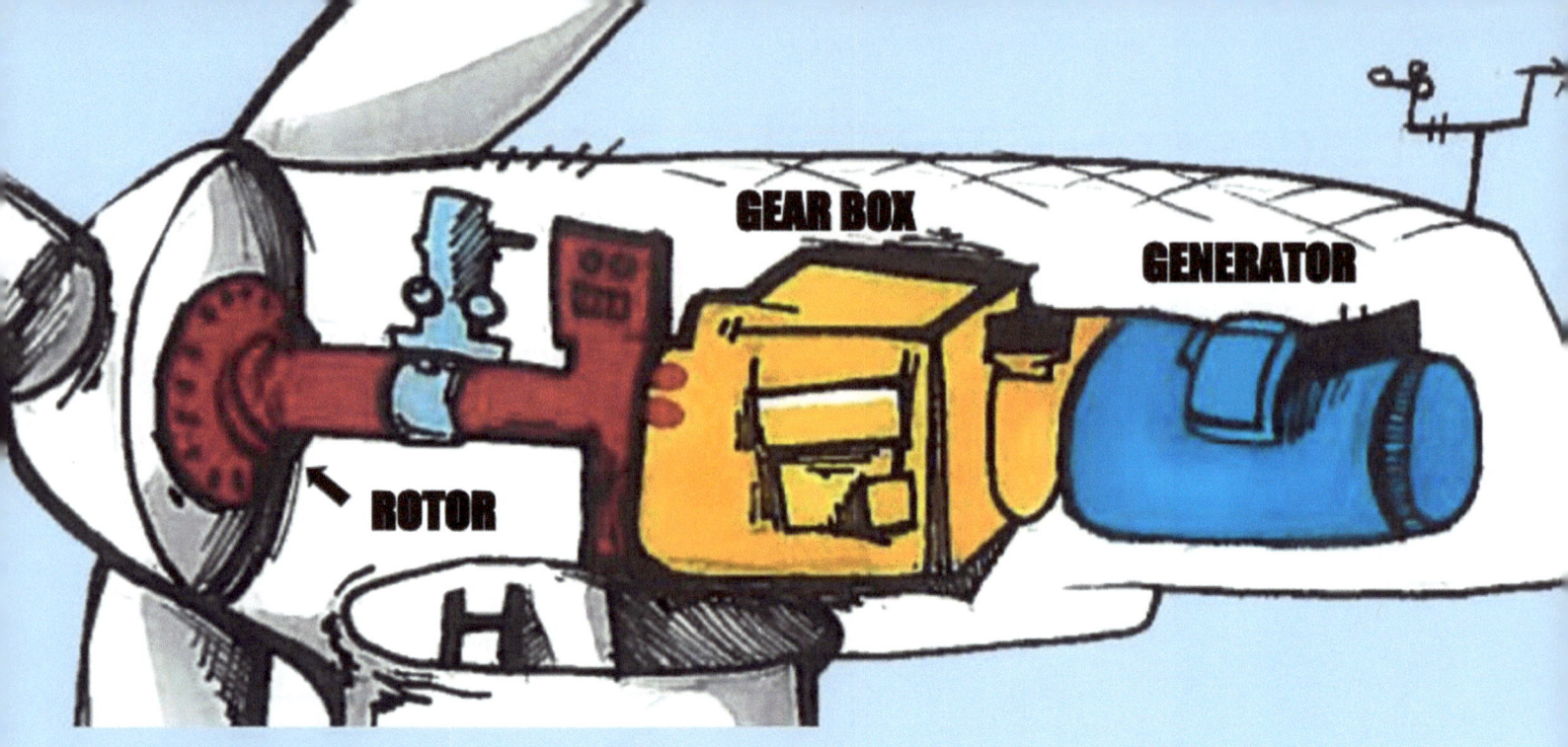

The electric generator converts the
movement of the gears into electrical
power by spinning magnets around a
wire to produce an electric current.
A wind turbine is different from a fossil
fuel power plant because it does not
need fuel to spin the magnets. Instead,
it uses wind! That is why it is called
clean energy.

The electricity travels through the tower into power lines to the substation. The substation then uses transformers to make the right voltage of electricity to travel into the electric grid and into the outlets in your home and school.

When Tad was finished talking, Beamer and John knew what they had to do! John decided to take action to help his new friends. He decided to get his mom involved. She convinced her company to ask American Wind Wildlife Institute (AWWI) for advice.

The researchers at AWWI listened to the problem and agreed that something had to be done. The wind turbines are important to creating clean energy, and the birds and bats are an important part of the ecosystem.

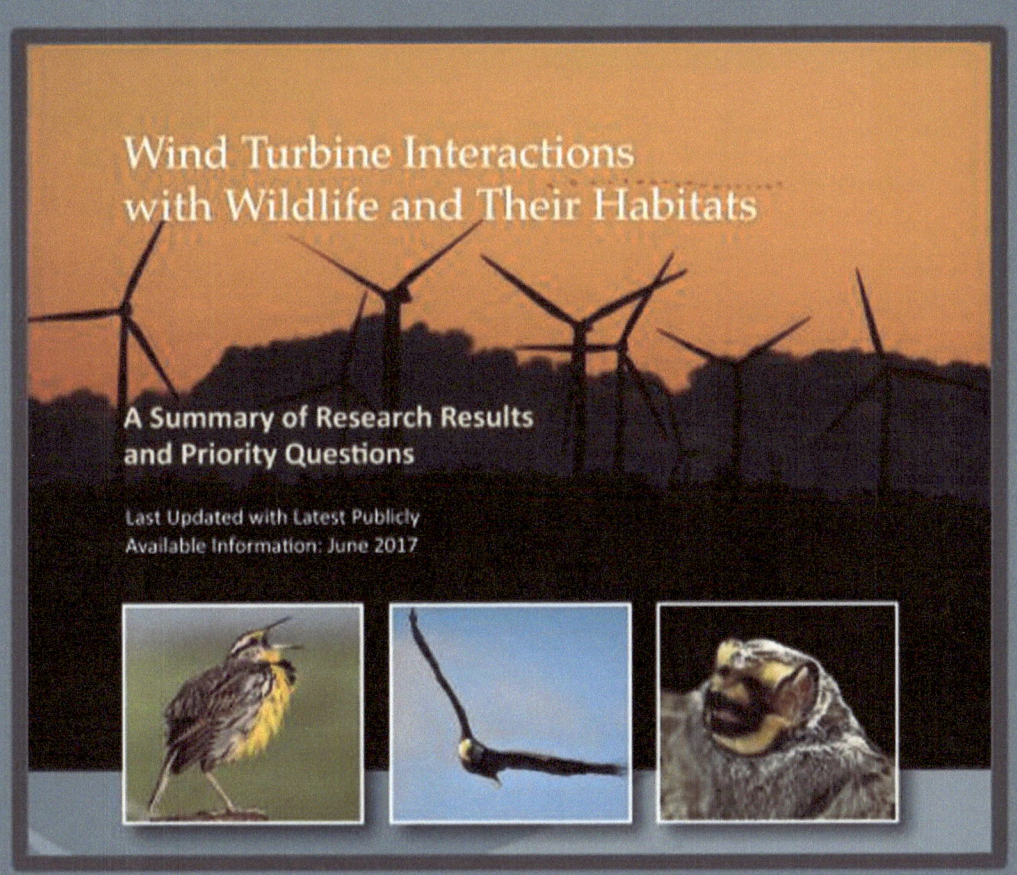

The scientists, Joe's mom, and her company decided they would install detectors to spot and warn approaching birds. The detectors could then turn off the wind turbines when Beamer's flying friends approached. The scientists decided they would test the systems with Joe's mom to see what worked best.

Because of John's hard work, he got to tour a wind farm, celebrate science, and help generate renewable energy, all while protecting his flying friends.

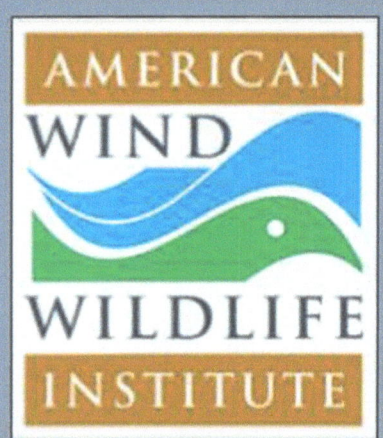

Profits from this book go to The American Wind Wildlife Institute. It is a non-profit, science-based, partnership of leaders in science and environmental organizations, the wind energy industry and wildlife management agencies who work together on a shared mission: To facilitate timely and responsible development of wind energy while protecting wildlife and wildlife habitat.

Combining the **Power of Science** with the **Voice of Collaboration** to facilitate Wind Energy Development and Wildlife Conservation

Seton LaSalle Catholic High School

Seton La Salle Catholic High school works hard everyday to give students a well rounded education that challenges them to ask difficult questions. It is through asking difficult questions that they come to a greater understanding of bigger issues in our world.

Grow a Generation

Grow a Generation partners with gifted and talented young people and teachers to make meaningful projects possible. Faculty, students, and student teams apply in their school to be accepted into the fellowship program. Once selected, they embark on a year-long odyssey to publish a book, create a digital artifact, or enter a STEM competition. Find out more at growageneration.com

Our Talented Illustrator

Sloane McCensky is a freshman at Seton LaSalle. She's been drawing all her life, using her free time playing Overwatch or Fortnite, drawing, or animating. She plans on majoring in animation and working with voice acting on the side.

Our Team Visit to the Patton Wind Farm

Our Authors at Work

www.ingramcontent.com/pod-product-compliance
Lightning Source LLC
Chambersburg PA
CBHW041302180526
45172CB00003B/935